Church House, Crowcombe - a history

FOREWORD

For many generations of villagers from many hundreds of English villages, the presence of a semi-public building in the shadow of the parish church was such a familiar sight that few ever thought to make a written record detailing how and why it was built, how it was used or what it meant. In materials and character, it would blend in with the village and yet, with its distinctive external staircase and first-floor entrance, it would stand out as something grander, a symbol of pride and prosperity. Scouring centuries of parish records even now reveals only oblique references; clues to materials and construction peek from entries for repairs; hints to usage hide behind payments for rent or hire; names such as 'church house', 'parish house' and 'community hall' reveal local preferences for nomenclature for that building.

Notwithstanding their antiquity, it is only in recent years that church houses are receiving scholarly attention, now combining archival research with architectural analysis to shine more light on those enigmatic documentary references and thus to set on firmer ground our understanding of their place and role in village life.

Hundreds have not made the long, tempestuous journey from their beginnings in the 15th and 16th centuries to today, and can now only be seen as a half-completed jigsaw puzzle in the imagination of the antiquary. Of those that have survived, scores more have lost their kitchens, their fireplaces, their halls and, much-altered, serve now as dwellings or public houses. However, a modest handful of these community buildings remain exactly as and where they have been for up to half a millennium; have seen-off or come through the changes time has wrought; retain their first-floor halls; and still stand grand and proud in the shadow of the parish church.

Church House at Crowcombe is one of these. From its beginnings as a community building in the village, it has been poor-house, school and ruin before returning full circle to where it began its journey nearly five-hundred years ago. A successful Heritage Lottery Fund bid permitted an extensive study of parish records to be undertaken - the final phase of a programme of grant funded external repairs. The combination has permitted an invaluable insight into the construction of Church House and its history over five centuries. Such detailed examination over a whole-life timeframe exists only for the smallest handful of extant church houses in England and so this narrative history represents an essential case-study to understanding them.

This history has been the result of inestimable hard work and dedication from the Church House charity, a group of local volunteers from the village. It is a great privilege to have assisted them, albeit in a truly small way, in bringing to life not only the history of this building, but also the history of the generations of villagers from this English village.

James Weir
Kellogg College
University of Oxford
December 2016

Church House, Crowcombe—An Overview

This beautiful, grade II* listed, 16th century building is one of Somerset's two surviving 'church houses'. Located in the heart of Crowcombe in West Somerset at the foot of the Quantock Hills, it has been witness to many phases of rural social life over the centuries.

Many of the church houses built around this time are in the south-west of England, and especially in Devon and Somerset. Some 53 church houses are known to have been built in Somerset, of which 21 have been identified. Only two continue to be used for their original community purposes. One is Crowcombe's Church House, and the other is in Chew Magna.

Originally built to hold 'church ales', Crowcombe's Church House has seen a variety of uses throughout its 500-year history. It has hosted a charity school, a poor house and a WWII soldiers' mess.

Early in the 20th century the building was structurally close to a ruin. The Rector of Crowcombe came to its rescue. He obtained charity status for it in 1907 and this led to a major restoration the following year.

Nowadays, Church House continues to fulfil its original role as a village hall for Crowcombe residents and the West Somerset community. It has been restored in more recent times too, having undergone major internal refurbishment in 2007 and external conservation in 2016.

Crowcombe's Church House is managed by volunteers from the village who report to the trustees. It is used for a range of community purposes, from family parties and wedding receptions, to club meetings and art exhibitions. It remains a much-loved building in our village.

Why was Church House built?

Up until the 15th century, the village church was the only place large enough to hold community events such as 'church ales'. These festivities, administered by the churchwardens, took place in the nave. Churches had yet to install pews and the nave was then an open area, separated from the sanctuary by a rood screen. It was the natural place for the village community to hold events, particularly in cold and inclement weather.

Church ales raised money for good causes such as repairs to the church and churchyard. They were often linked to specific dates in the church calendar—for example, Whitsun and the church's Patron Saints Day.

Ales were also an effective way of providing financial assistance to people in the local community. 'Bride ales' helped poor couples to fund their weddings, and 'clerk ales' enabled parish clerks to be paid for their work.

However, in the late 15th century the church authorities were becoming increasingly uncomfortable about alcohol consumption on church premises. Ales were often banished into the churchyard. Parishes were encouraged to build 'church houses' where secular communal events associated with the church could take place, while the church remained exclusively for worship. The churchwardens remained responsible for the management and use of the new secular building.

John Aubrey remembers (1714) the church ales held in his grandfather's time:

> There were no rates for the poor in my grandfather's days; but for Kington St Michael the church-ale at Whitsuntide did the business. In every parish is (or was) a church-house, to which belonged spits, crocks etc, utensils for dressing provision. Here the house-keepers met, and were merry, and gave their charity. (5)

Consequently, church houses, created specifically for the purpose of holding fundraising festivities, met the village's communal needs which the nave of the church had traditionally served. While worship and church ales were now physically separate, the church remained involved in the management of Church House—through its churchwardens.

Above: Church of the Holy Ghost, Crowcombe. The tower is the oldest part of the church and dates from the 14th Century. (Unknown source)

Opposite: Church House during the annual Flower Show. (Paul Savage)

The Origin of Crowcombe, its Manors and the Carew/Trollope Family

The Manor of Crowcombe has a fascinating history. In 1086, soon after the Norman Conquest, it passed to Robert of Mortain. From his 12th century descendants, the Crowcombe family then came to hold the land. (1)

In the 13th century the family set about increasing the profitability of the rural manor. In 1227, Henry III granted Godfrey de Craucombe [sic] the right to hold a weekly market. Rights to hold a three day fair followed in 1234, and later in the 13th century, the Manor obtained status as a borough. (1/4)

By 1236 the property and commercial rights of the Manor were shared between Godfrey de Crowcombe and Simon de Crowcombe. (2) Godfrey's share of the Manor was willed to the Prioress of Studley (a relative). Charitable gifts such as this, were common in medieval times for pious purposes. Reducing the time to be spent in purgatory was a powerful motivation for many. This part of the Manor became known as **Crowcombe-Studley** and was held by the priory until the Dissolution of the Monasteries (1539), nearly 25 years after Church House was built. (2)

Meanwhile, the other half of the Manor came to be known as the Manor of **Crowcombe-Biccombe**. In 1349, Simon de Crowcombe's great grandson, also called Simon, died childless. He was succeeded by his niece, Iseult, who married John Biccombe. Several generations later, in 1568, Elizabeth (a Biccombe daughter) inherited the estate. She was married to Thomas Carew of Camerton and the estate became known as **Crowcombe-Carew**.

A manor had existed at Crowcombe from the late 13th century. Five centuries later, Thomas Carew at the age of 21 began building the main house to plans by Nathaniel Ireson, who also designed Ven House near Sherborne. Carew completed the house by 1739. It remained in the family until the middle of the 20th century.

Despite the efforts of the family, their manorial ambitions for Crowcombe did not materialise. The number of houses in the town in 1791 was far fewer than had existed in earlier days. (3) The Carews' attempt to revive the market in the 1760s failed, in the face of competition from Bishops Lydeard. Ultimately, Crowcombe's market and fair were a shadow of their former selves and led to the parish reverting to a predominantly rural economy. (1)

A century later (in 1886) another descendant, EG Carew, died childless and the estate passed to his sister Ethel Mary who was married to Robert Cranmer Trollope. A remarkable transaction then occurred. In 1894, Ethel Trollope bought back the Crowcombe-Studley half of the estate and successfully reunited ownership of the entire original manor—after more than 600 years of separation. (2) When she died in 1934 she was succeeded by her grandsons AF Trollope-Bellew, who died on active service at El Alamein in 1942, and by Major TF Trollope-Bellew. Nowadays, the Major's son, Anthony Trollope-Bellew continues to own the estate.

For more on Crowcombe over the centuries, read A Short History of Crowcombe. (30)

CHURCH HOUSE: A GIFT FROM THE MANORS OF CROWCOMBE

Crowcombe's Church House came into existence through a gift in 1514 from the Lords of the Manors of Crowcombe-Biccombe and Crowcombe-Studley, Robert Biccombe and the Prioress of Studley, respectively. The gift comprised a house and garden opposite the churchyard, for the building of a 'church house' within four years.

The site of Church House is by no means an accident. Located within a stone's throw of the church and on the edge of the main thoroughfare, it was clearly intended to be a prominent village building and as accessible as possible. (4)

Its location demonstrates its link to the history of Crowcombe Court. Of special interest to historians is the triangle of land and widening of the road opposite Church House which almost certainly indicates a medieval market. Bordered on the north by Crowcombe Court and to the east by the churchyard wall, it is likely that this was the site of the Manors' weekly market, inaugurated in 1227 and

also used for the three-day fair. The entrance to the original Crowcombe Court was at that time almost certainly some 50 metres further up the road, at the end of the existing estate wall. (4)

It was during this time that Crowcombe became a borough and freeholders paid rent to the Lords of the Manor. The borough would have needed a building to manage its affairs and to hold a borough court. (4)

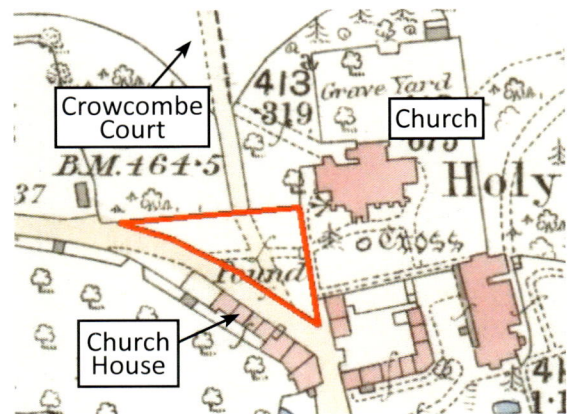

Left: Ordnance Survey (1888) showing probable site of medieval market in relation to Church House, the church and Crowcombe Court. (James Weir)

Above: The pound to the right of Church House was used to secure stray animals, animals grazing illegally on common lands or even to hold animals overnight for market. (Unknown source)

Top: Early sketch of Church House. (Unknown source)

Church Ales and the Reformation

Church House provided facilities to brew ale and bake bread, which villagers then consumed upstairs where they could dance and make merry. Downstairs there was an enormous open fireplace at the north-western end with two bread ovens. The food and ale were then carried in procession out along the front of the building (sheltered by a lean-to 'pentice' roof), and up into the gallery via the external steps. With no internal staircase, the pentice roof gave welcome protection from inclement weather.

Evidence of the pentice roof remains visible today. On the front elevation of Church House the horizontal band of protruding stonework once acted as the flashing for the roof. The roof was a canopy hanging out over the street, almost certainly supported on pillars.

A similar pentice roof can still be seen today at Widecombe-on-the-Moor in Devon. The canopy would have also provided shelter for the market traders, stalls and market officials. Church House's pentice roof is long gone. However, the position of the rafters can still be seen at roughly 1.8 m intervals along this front elevation, where square pieces of stone were inserted to replace the beams. Later, the first floor windows were enlarged and cut through the band of stonework. (4)

The building was originally thatched. This is demonstrated by the high level of the protruding stonework above the chimney stack to accommodate the thatch, and the shallower extension to the rafters required when slates took the place of the thatch.

Church House continued to fulfil its original community and church functions for some 150 years. But there were problems. At Somerset assizes in 1632, for example, there were even several indictments for the alleged murder of

Above: An early image of Church House detailing the external stairs and pentice roof line. (Unknown source)

Right: Front elevation of Church House showing the horizontal line of stonework where the pentice roof once ran along the length of the building. (Paul Savage)

Top right: A similar pentice roof and pillars at Widecombe-on-the-Moor in Devon. (James Weir)

illegitimate children conceived after church ales.

Later in the 17th century the Reformation gradually undermined the traditional role of church houses. Much stricter conventions on proper Christian behaviour became the order of the day. Secular amusements on Sundays were inadmissible. Church ales and alcohol consumption on Holy Days were an anathema for good Christian people. The role of church houses around the country now had to change to reflect the new religious and social climate.

In Crowcombe's case, its Church House evolved to fulfil two distinct roles for the community. Downstairs was converted to function as a poor house to accommodate the destitute, while a charity school opened upstairs to educate children.

Above: Church House was thatched originally. This is demonstrated by the chimney stack's protruding stone course starting at a higher level. (Unknown source)

THE POOR HOUSE

Well before the end of the 17th century the ground floor was adapted to house six poor people. This large space was divided into three separate rooms plus a kitchen. The two existing external doors gave access to one of these rooms and to the kitchen. Two additional doors were inserted at the east end of the building to give access to the other two rooms. The central rooms were heated by a double fireplace.

Six additional poor people lived in two adjacent semi-detached cottages that were connected to the eastern end of Church House by the external stairs—then covered by a roof. It is unknown when these cottages were built, but there is evidence of their maintenance by the churchwardens in 1789. The thatching of the cottages required regular maintenance, which became a routine expenditure item in the churchwardens' accounts:

> Paid Moses Gard 4/- for 16 sheafs of reed on Little House. [1789]
>
> 10/- for laying sheaves on the poor house. [1825]
>
> Paid 9/9 for thatching the Poor House. [1839] (9)

In these accounts the combination of the ground floor of Church House and the adjacent cottages were known as 'the poor houses', while upstairs was known as the 'school'. By the time of the 1842 Tithe Map, the two buildings were treated as one, namely 'No. 67'.

The churchwardens were responsible for the administration of the poor house, heating the rooms in the winter and financially supporting its occupants. The churchwardens' and overseers' accounts illustrate the kinds of work and support they provided over the years:

> 2/- laid out for the sheaves of wood in the cold weather for the Church House people [1695]
>
> 1/- paid for the same of wood for the poor people in Church House [1696]
>
> Paid Elizabeth Poole for cleansing out of the Church House after Sarah Pine died [1700]
>
> Paid Mary Oldman for tending of the people in Church House in their sickness [1723]
>
> Paid 3/1½ when the widow Radnage was brought to the Church House, for meat and drink, and putting up the bed. [1737]
>
> At the Vestry held at the School, 'tis ordered that John & Sarah Coles be permitted to live in the western part of the Church House. [1768]
>
> Paid John Coles to repair the roof of the school—1/5 [1784]
>
> Gave 5 men for extinguishing the fire in Poor House chimney—2/- [1825]

One intriguing meeting minute refers to having:

> Paid Messrs Rowcliffe & Son a bill for prosecuting William Herniman for selling Cyder during divine service—£10. 10. 8. [1846]

or £1,150 in 2016 money.

Top: The historic downstairs windows. (Catherine Brew)

Middle: The stunning original roof of the Gallery upstairs. (Paul Savage)

Bottom: Church House in winter. (Unknown source)

Left: The ground floor of Church House accommodated the poor. (Paul Savage)

Opposite page: Extracts from the Churchwarden's Accounts 1723 - 1807 detailing tiling, thatching and glazing undertaken on Church House. (Somerset Heritage Centre)

Churchwardens also took steps to ensure that the residents of Church House were sufficiently deserving to live there. From time to time it was necessary to require a family to leave—for example:

> It is ordered that James Burston, his wife and children, have Notice given them immediately to quit the Church House at Midsummer day next, he having sufficient to provide a House for himself and maintain his family. [1780] (9)

The social and economic upheavals of the 18th and early 19th centuries ultimately put an unbearable strain on the resources for poor relief. What transpired was the introduction of the drastic Poor Law Amendment Act of 1834. The Act compulsorily grouped parishes into unions for the construction and maintenance of workhouses. It split families up and deliberately made conditions in the workhouse worse than those of the lowest paid labourer living outside. It was a reform unpopular with the poor, but less so with ratepayers. (8)

Above: Williton, where the workhouse was located, is a village just over 5 miles north-west of Crowcombe. (Somerset History Centre)

Crowcombe joined the Williton Poor Law Union in 1836. Two years later in 1838, Williton Workhouse was opened and Crowcombe's able-bodied poor were transferred to the workhouse in the autumn of that year. A minute of the Williton Board of Governors required that:

> Parish Officers be required to remove Paupers in the existing Workhouses of the Parishes of Stogumber, Clatworthy and Crowcombe to the Union Workhouse on Saturday the 29th inst. [24 September 1838].

Williton Workhouse was built under the direction of Sir George Gilbert Scott, architect of St Pancras Hotel and the Albert Memorial in London. It was a grand building to look at, but conditions inside left a lot to be desired. Visits to workhouse residents were permitted, but were subject to strict rules:

> that the Relatives and friends of Paupers be admitted to visit them ... on each day in every week between such hours as the Visiting Committee may direct, it being understood that this order is not to restrain the Governor of the Workhouse from admitting at his discretion any friends or relatives to visit the Pauper inmates in cases of Sickness or Accident, nor to exclude any Persons authorised by Law for visiting the Workhouses as they may be entitled to do. [24 September 1838]

Now that the new Williton Workhouse accommodated the area's poor, Crowcombe had no need of a poor house of its own. However, in practice the cottages adjacent to Church House continued to be used as almshouses and to be administered by the parish.

Williton Workhouse: many of Church House's able-bodied poor were moved here. (Mary Evans/Peter Higginbotham Collection)

THE CHARITY SCHOOL

"... 15 poor boys of the Parish ...
18 poor girls of the Parish ..."

Whilst the poor house accommodated the poor downstairs, the Carew charity school educated children upstairs. The school is believed to have started in 1661 and is known to have been the beneficiary of three wills or endowments:

- 1669 – arising from the will of Elizabeth Carew
- 1716 – an endowment from the Rev Dr Henry James, son of a former rector
- 1733 – some generations later, from Thomas Carew's endowment (7)

In 1826 the Parish of Crowcombe gave evidence to the Royal Commission inquiring into Charities in England for the Education of the Poor. This gave a summary of the history of the Crowcombe Charity Schools and of the three bequests above.

It appears that the original source of funds from Elizabeth Carew in 1669 and the subsequent endowment from Thomas Carew in 1733 were for the education of boys. Crowcombe's evidence to the Royal Commission explained:

> The School established by Mr Thomas Carew has been in existence ever since, supported by its original endowments: 15 Poor Boys of the Parish being taught in it to read and write and cast accounts. The Master as well as the Boys are appointed by the Carew Family. The Boys have a suit of plain Clothes annually, and what remains after that expense is satisfied constitutes the Master's salary. The whole income of the School Property is about £40 a year. There is an old House which has belonged to the Parish before the Time of Memory on the ground floor of which, consisting of several apartments, the Parish Poor are lodged. In the room above, extending the whole length of the Building, the Children of Carews Charity School are Taught and the Sunday School is also held there.

By contrast, in 1716 Rev Henry James founded a Charity School for girls with his endowment. The evidence to the Royal Commission continued:

> 18 Poor Girls of the Parish are educated by two women between whom the income is divided. These girls are principally taught to read and write, and they as well as the Poor Boys above mentioned attend the Sunday School which is supported by Voluntary Contributions under the immediate superintendence of the Minister.

When Thomas Carew made his endowment in 1733, he also laid down very specific regulations for the charity school. Of the 40 children principally from the parishes of Crowcombe and Clatworthy, two thirds were to be boys and one third, girls. The charity school was to be:

> For the education of fforty [sic] Poor children ... of parents bred and educated in and propossing [sic] the protestant religion according to the usage of the Church of England by law established ... who are incapable of education ... out of any real or personal estate of their own.

His rules give insight into how the children were to be taught and expected to behave:

> The said children ... shall be in the school every

"... taught to read and write ..."

morning the Tenth day of March to the Tenth day of September at seven of the clock ... to Eight of the clock and the Schoolmaster shall attend at the said hours and ... the Lord's prayer and the Collect for the week and the first and second Collects set forth from the book of Common Prayer to be used at Morning Prayer, together with the general Thanksgiving and any other Prayers that shall be appointed by the rector of Crowcombe...

The said Schoolmaster shall have strict regard to the morals and behaviour of each child ... and he shall find any child guilty of lying prophane [sic] cursing and swearing blasphemous discourse ... or the said children shall not come in such cleanly and decent maner [sic] as the circumstances of the parents will admit, in every of those cases he do reprimand and correct the offender in a moderate and humane manner, and if the child ... shall not amend himself and behave himself as he ought to do, the Schoolmaster shall lay the matter before me—and after my decease before the rector of Crowcombe for the time being—in order that such Child may be displaced from the said school and some other child appointed in his room.

No child shall be taught to write before he can perfectly read the New Testament. (10)

When the Crowcombe School opened in 1872, the children being educated by the Carew Charity School at Church House were transferred to the village school. Church House was now empty and faced redundancy.

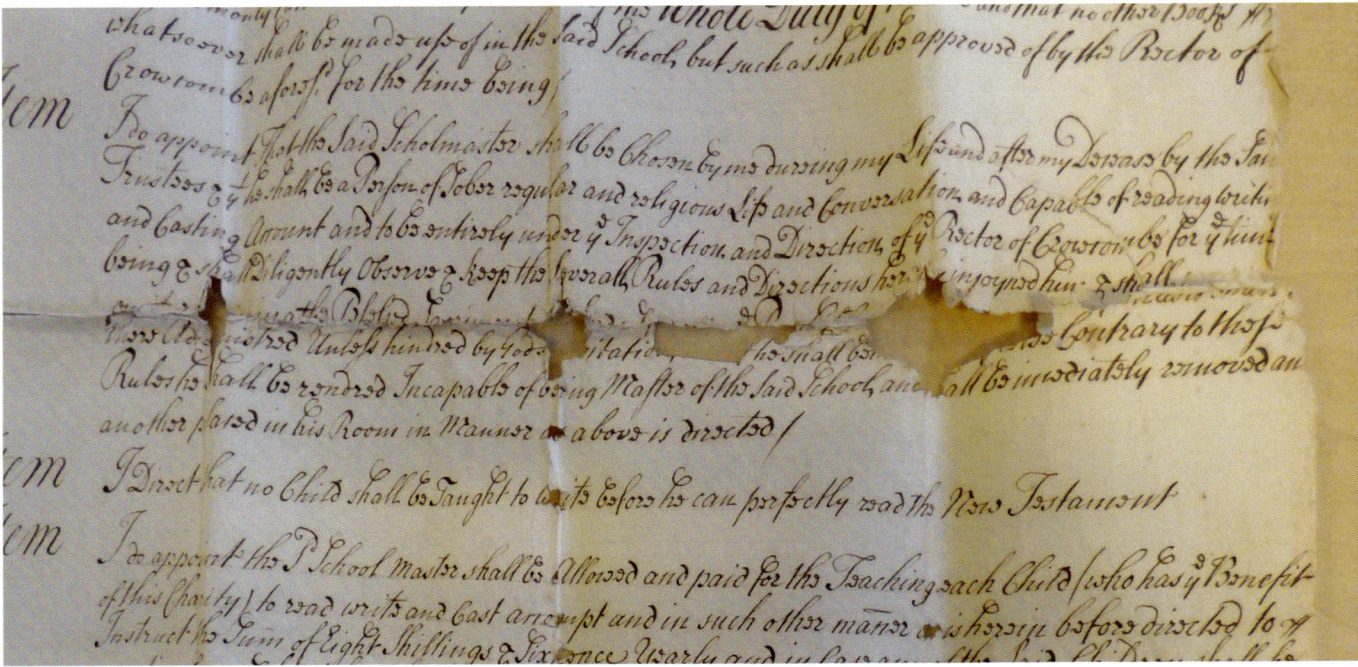

Above: An extract of Carew's Charity School rules from 1733. (Somerset Heritage Centre)

Church House's decline

The transfer of the poor to Williton Workhouse in 1838 had already begun to create financial problems for the parish. Despite pressure from the Williton Union, it was decided in 1841 at a key meeting of the vestry, not to sell the poor houses. However, there were still maintenance costs:

> At the request of the churchwardens, the state of repair of the front roof of Church House over the School Room, when it was agreed that they put it into proper repair on the lowest possible terms they can get it done for. [1841]

Paid mason's bill, £14.15. 9, for new slating the School House. [1842] (9)

Financial pressures must have been building up. In 1849 the vestry decided to sell the cottages and seek permission from the Poor Law Commissioners:

> for their consent to the sale … for the application of the produce thereof to the permanent advantage of the parish … of two freehold dwelling houses in the several occupations of Thomas Thresher, Henry Court and Betty Coles situate lying and being within the village and Parish of Crowcombe in the said County of Somerset and Union of Williton, subject to the payment of 8d payable to Mrs Mary Carew, Lady of the Manor, as an acknowledgement. (9)

Later in 1849, Mr TGW Carew's offer of £40 to purchase the cottages was accepted, but for some unknown reason this was not taken up. Instead, the cottages were sold at auction to a Mr Sweeting. That same year the remaining three occupants of the cottages were removed to Church House. In 1870 the vestry discussed the implications

Top: View from Church House to the church. (Catherine Brew)

Right: Crowcombe Vestry Book Minutes 1822-1845. Detail of the Vestry's decision in 1841 to not sell Church House. (Somerset Heritage Centre)

of the impending Elementary Education Act of 1870 and resolved:

> To ascertain how, if necessary, its requirements may be best met in the Parish ... Col Carew to write to the Charity Commissioners to ascertain their opinion with respect to the present schoolroom which is held by Trustees under them. (9)

Two years later in 1872 Crowcombe Village School opened, and the special role Church House had held in the village since the Reformation, had come to a close.

In anticipation of an empty building and little income, the vestry was faced with major financial problems. In 1876 a vestry meeting was called to discuss the need for repairs to the roof of Church House:

> For the purpose of consulting with the Churchwardens respecting repairing the roof of the old schoolroom ... (and) resolved unanimously that, if the Church House belongs to the Parish, the Parish should dispose of them and invest the money for the repairs of the Church. Resolved that Mrs Carew be requested to write to Mr Easton (representing the Carew estate) and Mr Wills (representing Sir RB Harvey) to ascertain whether they laid any claim to the said Church House, on behalf of their respective clients. (9)

However, the parish's discussions were only to become more complex. While they contemplated selling Church House, they realised that there was no formal documentation as to who actually owned the building. As the *Somerset County Gazette* later reported:

> Between 1870 and 1876 proposals were made to dispose of the property and invest the money for repair of the church, but it could not be ascertained that the Vestry had to right to sell it, while no-one seemed to know who were the actual owners. So the matter was adjourned indefinitely. (13)

However, in 1876, the Manors replied unequivocally that they were the owners of Church House:

> Respecting the adjourned question of selling the Church Houses, Mr Easton (the solicitor to the Carew Estate being present) stated that the Property was claimed (subject to the same being used for school purposes and the holding of Manor Courts) by Sir Robert Harvey and the Carew Trustees, and that they objected to any sale or disposition of such property by the Parish. [1876] (9)

Above: An early view of Church House. (Unknown source)

The opening of Crowcombe Village School, 1872. (Unknown source)

The vestry was not pleased and a motion was put:

> that the Parish do not repair the Church Houses, but that they be given up to the Lords of the Manors of Crowcombe Studley and Crowcombe Biccombe. (9)

However, cooler heads won the day. The minutes record:

> An amendment to this motion was carried by 7 : 3 majority that further consideration of the Church House property and any repairs thereto to be adjourned for further consideration. [1876]

In practice, there was never any maintenance done on the roof. (9) A tentative ambition to restore Church House arose in 1897. At a parish meeting:

> Upon the occasion of the late Queen Victoria's Jubilee [1897] a suggestion was made that a restored Church House would be a fitting memorial of this auspicious event. (7)

Again, nothing came of this proposal. By 1897 Church House was in a ruinous state. Too many years of indecision and inaction had taken its toll.

We have little detail about the use of Church House for some 35 years after Crowcombe Village School opened in 1872. Upstairs continued as a Parish Hall, and downstairs was converted into a Men's Institute. By 1887, the pauper's room at the east end of the building had been converted into a reading room. The partition between the other two poor rooms seems to have been removed to become a Billiard Room. Our evidence is a remarkable combined barometer/thermometer/calendar that was presented to Charles Jordan by his fellow members in 1906 'on his leaving Crowcombe after 19 years Hon Sec of the Reading Room'. (11)

He had been the village's shopkeeper since 1871. Towards the end of his 35 years of serving the village community, his business had an annual turnover close to £900 (nearly £100,000 in today's money)—a significant rural enterprise. (11)

However, towards the end of his 19 years as Honorary Secretary, it cannot have been much more than a nominal role. The roof of Church House had collapsed, as had the floor of the gallery.

Above: The barometer presented in 1906 to Charles Jordan, the Hon Sec of the Reading Room. (Valori Menneer)

The ruinous building, 1907

Various images of Church House in a badly delapidated state, 1907. Both the south facing roof and the upstairs floor had collapsed. Vegetation was growing from the roof and the building was close to ruin.

Above: (Unknown source)
Far right: (Unknown source)
Right: (Nooks and Corners of Old England)

The restoration of Church House

The Rector of Crowcombe, the Rev Henry Christian Young, came to the rescue of Church House in 1907.

He had been educated at Bradfield School and at Jesus College, Cambridge. He was a bachelor, living with his widowed mother, four sisters and a brother at Crowcombe House, the principal dwelling in Crowcombe—other than Crowcombe Court itself. He had been appointed Rector of Crowcombe in 1901 at the age of 35. He went on to serve the parish as its rector for over 40 years, dying in harness in 1943. (12)

Three years after his appointment, aided and abetted by his clerical colleagues, the Rural Dean (Dr Powell) and the Rector of Dodington (the Rev William Greswell), he approached the Charity Commission to ask whether Church House could become a charity. A formal charitable status would enable Church House to engage in fundraising to restore the building.

In response to the rector's enquiry, the Charity Commissioners checked with Mrs Ethel Trollope whether the estate had any claims over Church House. Two days after Christmas 1905 she replied unequivocally:

> Your Commissioners are evidently mistaking the Old Church House (which is still existing although in ruins and part of the Manors of Crowcombe Studeley and Crowcombe Bickham) for the Old Parish House which was sold by the Parish at the time of the Poor Law Unions to a Mr Sweeting ... The Ruins of the Church House form part of the Manors and it was in this House the poor nominated by the Rector and Churchwardens were received and the Carew Charity held a School previously to the building of the present school. (14)

Mrs Trollope's contention galvanised the rector into energetically digging through the archives. In early 1906 he set out to demonstrate to the Charity Commission that the Manors had made a gift of Church House to the parish, and that over the centuries the churchwardens had administered Church House on behalf of the community.

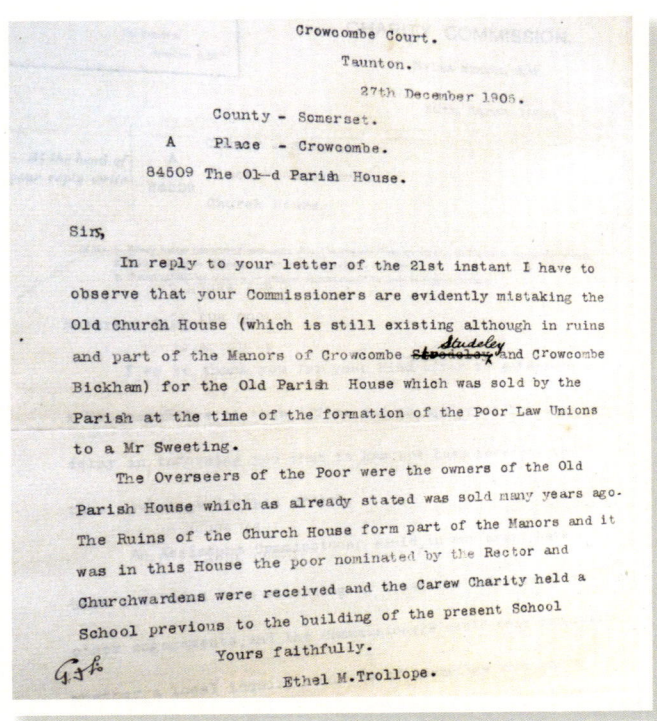

Top: Rev Henry Christian Young. (Unknown source)

Above: Mrs Trollope's letter to the Charity Commission regarding the Estate's claims over Church House. (Somerset Heritage Centre)

The rector drew attention to a memorandum dated 1820 in the 'Crowcombe Poor Account':

> Property belonging to the parish of Crowcombe for the use of the Poor. A poor House and two cottages adjourning in fee subject to a Lord's Rent to G H Carew Esq of 4d per year and also a Lord's Rent to Robert Harvey of 4d per year. (15)

His letter continued:

> Here is an acknowledgement that the Parish were owners not only of the two cottages (which were sold, and which Mrs Trollope calls "the Parish House"), but also of the "Poor House" that is "the Church House".
>
> The Lord's Rent has been paid annually from 1723 (the earliest date of Churchwarden's Accounts still existing) up to the present time by the Churchwardens, the last payment being made Easter 1905 and consists legally of 4d a year to each Lord of the Manor...
>
> This rent does not affect ownership, but is only acknowledgement to the Lord of the Manor dating from feudal times. By an act known as 31 Henry VIII.ch.13 the Rents and Services of the Lords of Manors of confiscated Religious Houses went by right to the King. (15)

He concluded:

> The above evidence reveals a very strong case in favour of the Church House being the property of the Parish, and I believe is too strong to be overcome by the mere statement of the present Lady of the Manor, which statement up to the present has never been supported by any evidence whatever.
>
> The claim by the Lord of the Manor was first brought forward in 1870 when the new schools were built, and the disposal of the old school came under consideration. The matter was allowed to drop, and since neither the Lord of the Manor nor the Parish have taken any steps to settle the question, and neither party has done any repairs to the building, it is now in a very ruined state. If the Lord of the Manor had evidence in support of his claim, why has he not made good his claim, instead of allowing the building to become a perpetual eyesore, exactly opposite the entrance to his house? (15)

Above: An extract of the rector's letter to the Charity Commission. (Somerset Heritage Centre)

With such clear archival evidence that included details of payments by the churchwardens and overseers on the maintenance of Church House over the centuries, the rector must have been hopeful of his desired outcome. Nonetheless, he no doubt much appreciated the encouragement from the Rev. William Greswell who wrote to him in support:

> Just to say that I hope you <u>will not lose</u> any time in stirring up the Charity Commissioners. Would the Court people ever think of pulling down the Church House in question? (16)

Much to the rector's delight, in February 1907, the Charity Commission accepted Church House as a Charity, with the trustees to be:

- The Rector and Churchwardens of the Parish of Crowcombe
- The Rural Dean of the Rural Deanery in which the Parish of Crowcombe is for the time being comprised
- The Archdeacon of Taunton

The Commission required that:

> The building belonging to the Charity shall be used for parochial purposes in connexion with the Church of England as the Trustees from time to time think proper. (17)

Above: Confirmation of Charity status from the Charity Commission. (Somerset Heritage Centre)

Above: In ruins. (Unknown source)

Although Charity status had been awarded, the rector remained concerned about Mrs Trollope's ownership contentions, until the Charity Commission informed him three months later that:

> The Commissioners have now been informed by Mrs Trollope that she does not intend to proceed further with her claim to the ownership of Church House. (18)

The architect's findings

The rector now got to work and appointed an architect, Mr A Basil Cottam. He first asked for an initial opinion and advice on the structural state of the building. Mr Cottam's report on Church House was unequivocal:

> It is extremely dilapidated and rapidly becoming ruinous. Prompt measures should be taken if it is to be saved from serious structural failure.
>
> As regards the main walls, which are built of rubble stonework and are about 2ft thick, I find that the North wall, facing the road, is decidedly bulged outwards and is not strictly perpendicular in any part. The South wall is also out of perpendicular and is inclining inwards. It is buttressed on the inside by a very large chimney stack which, at present, gives it great support and prevents further inward movement. The East end wall appears to be fairly sound. The West end wall is in the worst condition of all. It is very badly fractured and to a large extent needs rebuilding. The parapet walls are ruinous and their coping has so far disappeared that I fear it may be impossible to ascertain what originally covered them.
>
> All the walls are suffering from the ravages of weather, which has washed out the mortar joints and damaged the stonework. In places ivy and other large growths have obtained a strong hold in and over the walls, doing great mischief. The old First Floor has very largely disappeared and the "common" joists which carried it are mostly gone. Those which remain are inadequate and are failing under the influence of age and weather.
>
> The roof is an exceedingly graceful one typical of the date at which it was put on. It has seven framed trusses with cambered collars, high up, and arched ribs, carrying purlins, placed on the flat with curved wind braces under them. The common rafters are laid flat and are covered with small rough slates. Most unfortunately nearly the whole of the South side of this roof has been allowed to fall

"It is extremely dilapidated and rapidly becoming ruinous. Prompt measures should be taken if it is to be saved from serious structural failure."

in and very sad damage has, in consequence, resulted to it and to the rest of the Building which has, by this, been laid open to the weather.

Dealing with the restoration of the Building and fitting it for use in connection with the Parish to which it belongs, I have to report my opinion that, with care, it may be again be rendered perfectly sound and fit for Parochial purposes without destroying its character.

I think it may safely be assumed that an outlay of certainly £500 will necessarily be involved and that the details of the scheme, when settled upon, may easily extend this sum in amount. (19)

and adaptation, for submission to the Trustees...

Subject to the weather being favourable, I will arrange to come over for the purpose on Wednesday next the 29th inst, with an assistant, by the train due at Crowcombe Station at 11.03am, when I shall be greatly obliged if you can conveniently meet us and drive us from the Station. I am much obliged for your kind invitation to Lunch on that occasion, of which I will gratefully avail myself. I am, Dear Sir, yours truly... (20)

Above: Church House before the 1908 restoration. (Unknown source)

Below: Cottam's hand-written letter to the Rev H Christian Young. It begins with 'I beg to acknowledge ...' (Somerset Heritage Centre)

Planning the restoration

Mr Cottam was then invited by the Church House Trustees to undertake a survey of the building and to prepare drawings, specification and cost estimates for its restoration. He was invited by the Rev. Young to attend a briefing meeting with him—and to stay for lunch. He replied:

> I beg to acknowledge receipt of your letter of the 22nd informing me of the recent meeting of the Trustees of the above Charity and forwarding their invitation to take a complete Survey of the buildings and to prepare Drawings, Specifications and Estimates for the proposed works of reparation

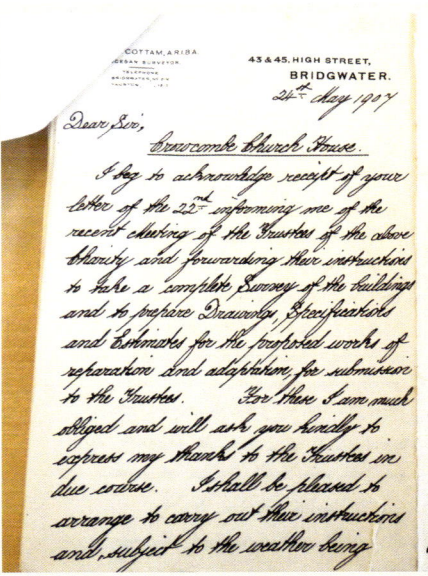
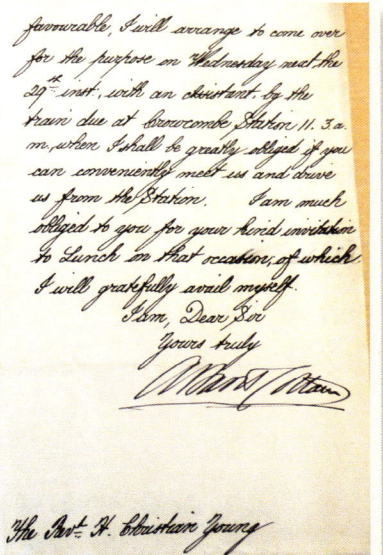

While the Manor appeared not to claim any ownership of Church House, the Lord of the Manor continued to present some day-to-day logistical difficulties. Mr Cottam wrote to the rector:

> I am surprised at Mr Trollope's refusal to let us go into the Orchard. I do not for a moment believe that he can prevent us from going there for legitimate purposes in connection with the building ... I think your view is probably correct and sound and that we should simply go there and do what is necessary, taking care to cause no damage to his property. (21)

Despite problems with access, the architect was able to prepare his findings and began briefing the Charity Commission on his restoration plans:

> The building shall be rendered useful for parochial purposes, as far as can be, without undue alteration. You mentioned you wish that it should contain on the Ground Floor a Reading Room, Games Room, a Kitchen and Store with access to the First Floor, and that the latter should be devoted to the purposes of a Parish Hall...
>
> The only structural alteration will be in connection with the Reading Room. In order to make this a fair size, it will be necessary to move the old cross wall (which I am not at all sure is an original wall) slightly further to the West, and to insert a small window in the North wall in place of the door which has been knocked through during some former alteration...
>
> The two ancient doorways in the North wall are used, one for approaching the Games Room and Reading Room, the other for access to the Hall and Kitchen ... I hope to be able to reuse one of the ancient doorways now in the Building between these two Rooms...
>
> In order to better the light in [the Games Room], I propose restoring the ancient window which has been knocked away to form a door at the North East end of this room ...
>
> Out of [the Kitchen] a Staircase would rise to the West end of the Hall. I suggest that the principal approach to the Hall should be up the existing ancient flight of external steps, at the east end, and through the doorway which already stands in position here ...
>
> As regards the works of reparation:
>
> 1. <u>Roof</u> ... lay bare main timbers ... reusing the sound old slates.
>
> 2. <u>Walls</u> ... North wall requires little done to it ... The South wall will require to be taken down in part ... The large chimney stack in the centre should also be taken down and reconstructing ... for the purpose of warming the Games Room on the Ground Floor and the Hall above. The existing chimney stack in the South East corner of the Building, which is clearly an insertion, should also be reconstructed and used in connection with the Reading Room. The two ancient windows in the South wall, now blocked up, should be reopened.
>
> The East wall requires but little work done to it beyond the gable coping which must be renewed having nearly all disappeared.
>
> The West wall is very shaky, has settled and cracked and, having been a deal pulled about, is decidedly unsatisfactory. It is also of great thickness. I am inclined to advise that it be largely taken down and, if possible, constructed rather less thick than at present, at least as far as the springing of the roof, above which point

the present thickness of wall might be retained. The present fireplace in this wall should be reconstructed for use as a Kitchen fireplace, and the inserted fireplace in the South West corner should be done away with...

I am of opinion that the cost of doing what I propose will amount to £550. (22)

Above: Cottam's pre 1907 ground floor plan of Church House, before any restoration works had been undertaken. (Somerset Heritage Centre)

Below: Cottam's pre 1907 first floor plan of Church House, before any restoration works had been undertaken. (Somerset Heritage Centre)

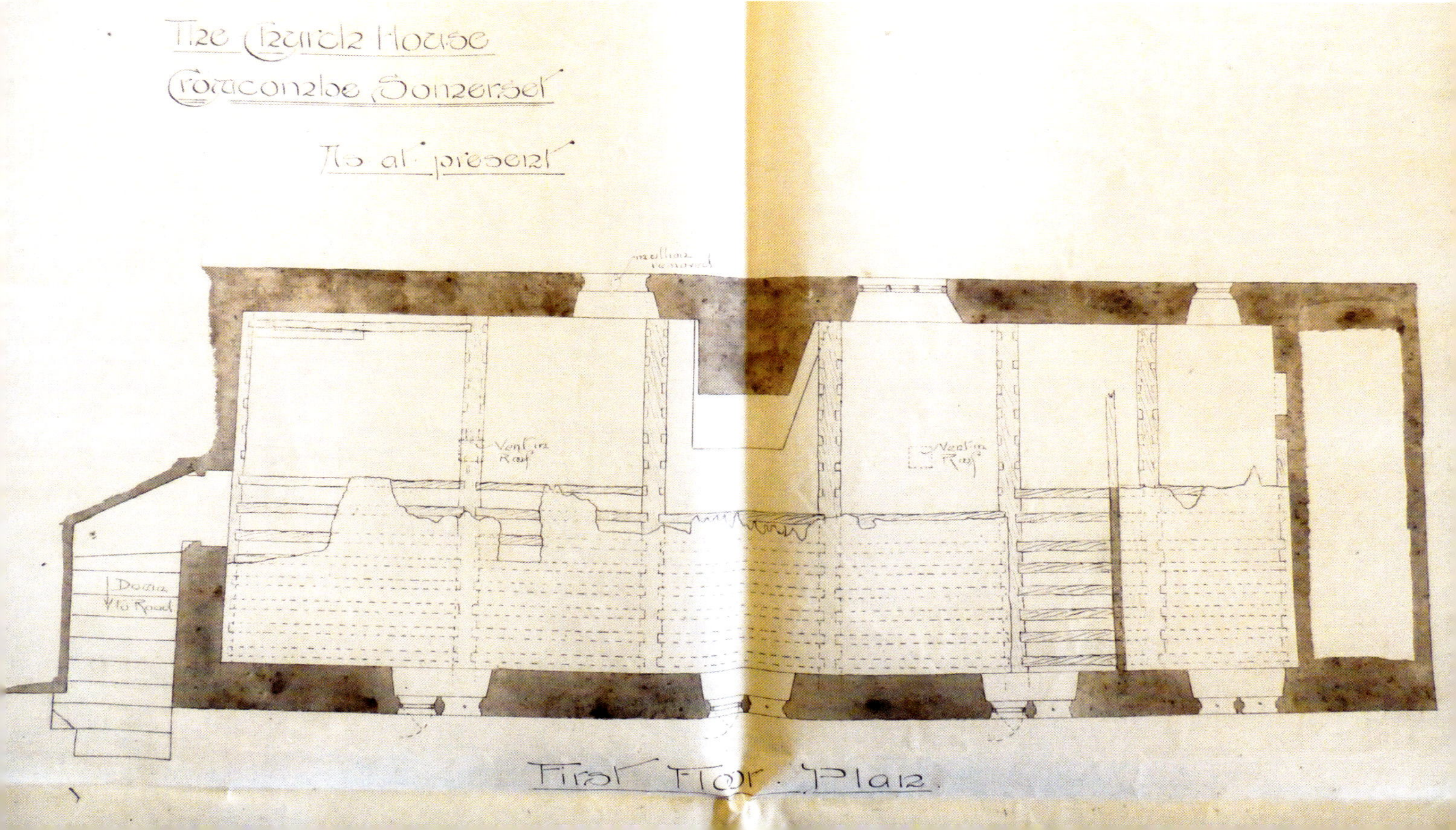

✠ The Church House, ✠
CROWCOMBE.

An Appeal for Funds.

THIS fine old building, situated opposite the Parish Church, and called "The Church House," is of great historical and architectural interest.

Its history can be traced back to Pre-Reformation days. In the year 1515, it was presented to the Parish by the then Lords of the Manors of Crowcombe Biccombe and Crowcombe Studeley. Collinson, in his "History of Somerset," states that "Robert Biccombe 6 Hen. VIII. made a grant of his moiety of the Church House (the Prioress of Studeley at the same time giving up her moiety) towards the repairs of the Parish Church of Crowcombe."

In "Lord Brougham's Commission concerning Charities" (1819-1837) we read, under the heading "Crowcombe":—"There is an old house which has belonged to the Parish before the time of memory; on the ground floor of which, consisting of several compartments, the parish poor are lodged; in the room above, extending the whole length of the building, the children of Carew's Charity School are taught, and the Sunday-school is also held there."

The walls of the Church House, which is over 60ft. in length, are built of rubble stonework, and are about 2ft. thick; and the roof is an exceedingly

While Mr Cottam had been making good progress surveying the building and in his planning, the restoration was not going to happen without funds. In the summer of 1907, the rector began fundraising in earnest.

The parish minutes record:

> A Meeting of the Parish was held at the schools on June 12th when it was unanimously resolved that an effort should be made to raise by subscriptions the sum of £500 to restore the ancient Church House, and to adapt it for the purposes of a Parish Room and a Men's Institute. (13)

Bottom left: Minute of Parish Meeting recording decision to raise £500 by subscription. (Somerset Heritage Centre)

Bottom right: The Parish sought subscriptions as a means of raising the much needed funds. (Somerset Heritage Centre)

Below: Detail of the list of subscribers. (Somerset Heritage Centre)

Opposite page: The rector's 1907 appeal for funds. (Somerset Heritage Centre)

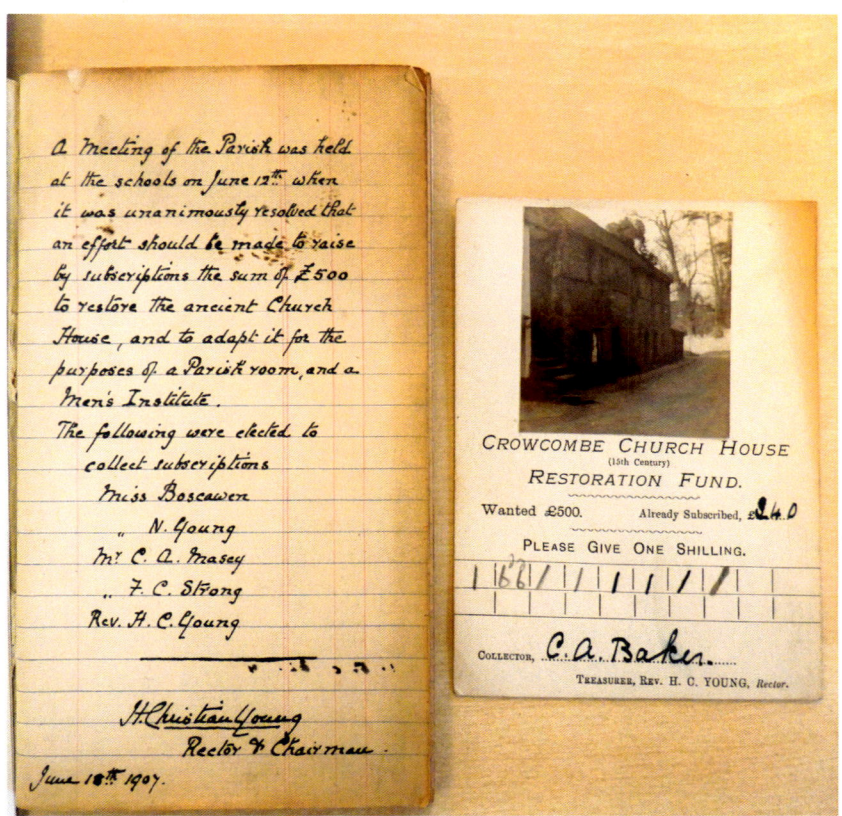

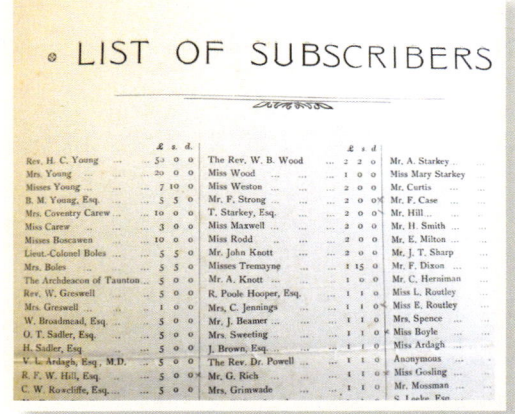

Restoration

Mr Cottam's comprehensive plans were well-received and at the end of September 1907, the Charity Commission approved his proposals. Work was already underway two months later, in November.

Unfortunately, the west wall was found to be in an even worse structural state—and Mr Trollope continued to present obstacles to the enterprise, as Mr Cottam reported to the rector:

> Mr Trollope turned my assistant out of the Pound yesterday. I suppose he had a right to do so. It will be very awkward if he keeps us away from this side of the building. I send you a photograph of the West wall. The further we look into this part, the worse it appears. I have taken a record of the individual stones to guide us in restoring it and now ordering Spear (the building contractor) to take down the shaky parts, numbering the stones as he does so, and to re-erect it all as before. (23)

One of Mr Cottam's key innovations was the construction of internal stairs from the kitchen area up to the first floor. However, he proposed retaining the existing external steps as the principal approach to the first floor. The proposals he sent to the Charity Commission envisaged:

> Out of [the Kitchen] a Staircase would rise to the West end of the Hall. I suggest that that the principal approach to the Hall should be up the existing ancient flight of steps at the east end and through the doorway which already stands in position here. (22)

The plan to install an internal staircase became more complex. He was required to construct a removable platform at the west end of the hall, presumably as

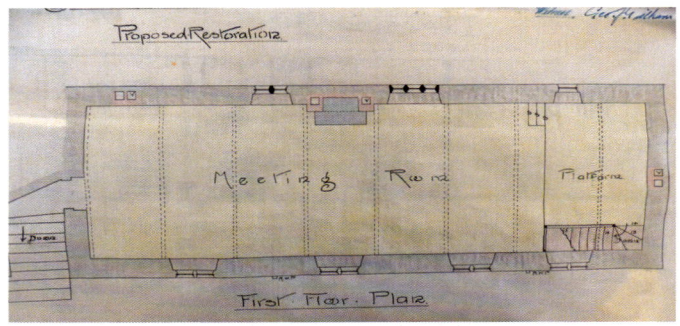

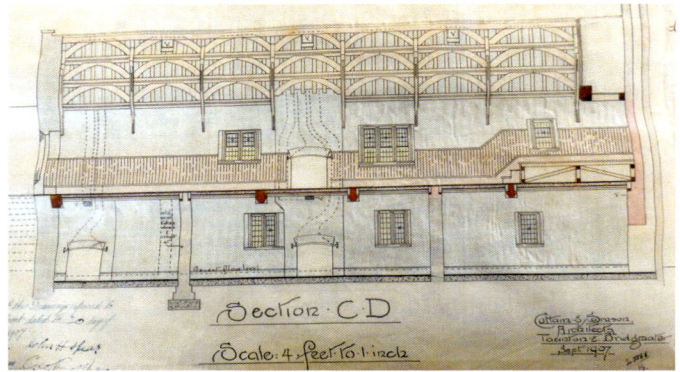

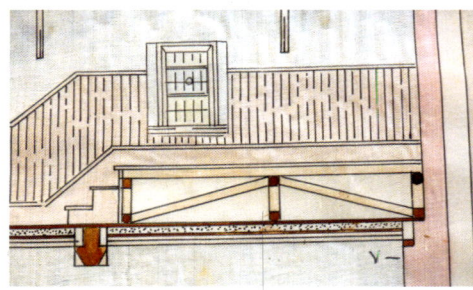

Top: Cottam's proposal for the first floor. (Somerset Heritage Centre)

Middle: Cottam's proposal included a removable platform. (Somerset Heritage Centre)

Bottom: Detail of the platform. (Somerset Heritage Centre)

a stage for concerts and theatrical performances. Mr Cottam sought a clearer brief from his client:

> May I ask you kindly to let me have your views and wishes with regard to the platform at the West end of the First Floor of the building. My chief reason for asking is that I have now to arrange the stairs from the Ground Floor to the First. It will simplify matters for me if the stairs have only to reach from floor to floor, and not also to be carried on to the extra height of the platform. I think I gathered that the platform was to be a moveable one, so that the floor might, if necessary, be cleared for occasions such as dances etc. (24)

Mr Cottam then came forward with his final proposal:

> I propose that the platform should be made of trestles, on which rather heavy planks of, say, 9" x 2½" should be laid ... very easy to move away and store when you wish to clear the floor.
>
> The steps to the platform would form a continuance, in a direct line of the stairs, as they take a quarter turn at the top and these steps would be removable along with the platform. (25)

In reality, the design of the staircase must have posed considerable problems for villagers serving as kitchen staff. Firstly, it was very steep, rising at an angle of approaching 45 degrees. Secondly, when the staircase was not being used, the closed cover served as the upstairs floor. This hatch cover was 1.8m by 0.9m and hinged on the road side. It was opened by pulling on a cord suspended from a swivel attached to the beam above. Any villager carrying a tray of drinks up these steep stairs somehow had to open the hatch before emerging into the upstairs gallery.

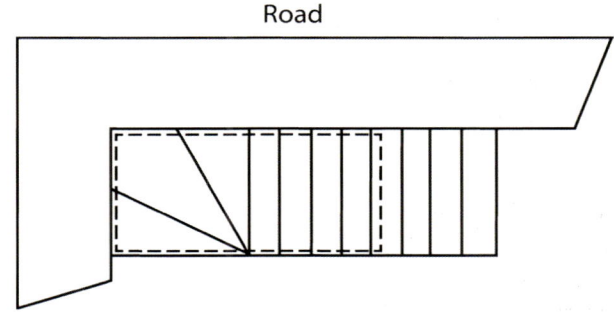

Top: Cottam's first floor with the platform at the far end of the room. (Unknown source)

Above: The 1908 internal staircase. The dotted line shows the hatch cover. It was hinged on the road side of the building and was level with first floor.

Restoration Extras

The final restoration included a number of features that had not been envisaged at the outset. We can see from Mr Cottam's final invoice submitted to the trustees, that these additional items included:

> Additional work to securing and casing West wall.
> Providing and setting furnace and carrying up additional flue
> A kitchen range
> Laying on water from main to Kitchen.

Presumably, up until then water had been drawn from a well, but there is no evidence for this.

Mr Cottam's invoice continues:

> Supplying and fixing rainwater gutter to South wall ... drain under [sic] building to road and connection to sewer (drain bedded and surrounded in concrete).

No doubt this layout was a less than convenient arrangement, but was necessary to avoid more conflict with Mr Trollope who owned the orchard immediately to the rear of the building. A conventional soakaway would have required intrusion into Mr Trollope's orchard. This drain was inserted into and across the floor of Church House to connect up with the sewer at the front of the building. It survives to the present day.

Mr Cottam's last addition to his invoice refers to: *'Pointing outer face of North wall'*. The front of the building was the only elevation to have been systematically re-pointed in 1908. (26)

Despite the unplanned additions to the restoration project, the final cost turned out to be precisely the figure that Mr Cottam had forecast at the detailed planning stage a year earlier: £550, or little more than about £55,000 today. (26)

Top: Church House. (Valori Menneer)

Bottom: The arch-braced purlin roof. (Paul Savage)

Restoration comes to an end

Within a short space of time and after a lot of hard work, Church House had been reborne from a ruin to a beautiful fully restored building. All in all, this extensive restoration included:

- re-roofing of the entire building
- construction of a new floor for the gallery
- windows to replace the two external doors
- major rebuilding of the north-west wall
- raising the original fireplace lintel to its present position and
- the installation of a steep internal staircase with its hatch cover

Celebrations and a grand opening

In September 1908, the village celebrated the building's restoration with a grand re-opening ceremony, preceded by a service of thanksgiving in the church (a shortened Evensong). By all accounts it was a magnificent occasion. Both the *West Somerset Free Press* and the *Somerset Gazette* gave lengthy accounts of the proceedings with verbatim reporting of the sermon and speeches given by the Archdeacon of Taunton and the Rural Dean:

> The Archdeacon, in calling upon Dr Powell to move a vote of thanks to the Hon Mrs Trollope said she had come there at great inconvenience, having come down from London expressly for the purpose of opening the building … They also owed a debt of gratitude to the Rector for all that he had done (applause).
>
> Dr. Powell congratulated the parish on possessing such a useful and delightful Church House which had been restored in the best manner possible. They were going to start it well with a concert that afternoon and social festivities in the evening, and he hoped that it would go on and prosper. (27)

After almost losing the building to a ruin, it was clearly important to the village that Church House had survived. In his speech, Dr Powell explained the Charity's Trustee structure, highlighting the importance of Crowcombe

Top: The rear of Church House, n.d. (Unknown source)

Bottom: The Somerset Gazette's report about Church House's grand opening. (Somerset Heritage Centre)

members having a majority amongst the trustees:

> Referring to the difficulties that had arisen, the greatest difficulty, he pointed out, had been the making of a scheme of trusteeship, but five trustees were now appointed of whom three must be parishioners—the rector and two churchwardens—but as it was possible that the persons filling those offices might not always take such an interest in the building as the present rector and wardens did, two outside trustees were provided in the Archdeacon of Taunton and the rural dean, but he pointed out that so long as the parishioners took keen interest in the building outsiders could not interfere because the parishioners had the majority. (27)

Following a contribution from the rector:

> The Rev H C Young observed that as two members of the Carew family founded a school in that parish it seemed to him that Mrs Trollope was the most fitting person to re-open the old church house, and so carry on the traditions of the past. (Hear, hear). He was immensely grateful for the generosity of the whole neighbourhood of West Somerset, as it would have been impossible to have restored that building in the way it had been done without that kindly help. Today he could tell them that he had raised over £400, and it would not be long before he obtained the other £150. (Applause)

Mrs Trollope was then invited formally to re-open the building:

> The Hon Mrs Trollope, in acknowledgement, said that not in his wildest dreams could her ancestor, Godfrey de Crowcombe, ever have thought that the parish church house would be put to use in the twentieth century. Without doubt a building of that character might be of the greatest benefit to any parish or community. (28)

Consistent with Church House's original purpose of hosting village festivities, the formal part of this 1908 grand re-opening ended with a concert that merged into a village party—with dancing in the gallery continuing into the small hours. It must have been rewarding for everyone involved in the restoration to be able to enjoy the beautiful building once again.

Today, over 100 years later, Church House continues to retain its original function as a village hall. It remains a much-loved building at the heart of the village.

Church House: from the 1908 restoration to present day

From 1908 onwards, Church House has continued to serve the village and the wider West Somerset community in many ways. Its uses have responded to the needs of the day.

Within a month of WWI being declared, a village meeting was held in Church House in response to Kitchener's Call to Arms, to recruit and enlist volunteers in the Somerset Light Infantry Regiment. On 5 September 1914, the *West Free Press* reported:

> A meeting was held in the Church House under the presidency of Mr F H Cheetham who was supported by the Hon Mrs Trollope, the Rev H C Young and other influential residents. Appeals for recruits were made by the chairman, Lieut Goodland and Colour Sgt Davis, with the result that the following seven in all gave in their names: William Walsh, Percy Duddridge, C Grant, G Radford, Bert Durrant, Frederick Smith and William Calloway. Mrs Trollope afterwards generously announced that any young man in her employ who desired to join would be paid a half wage in service, and their posts would be kept open for them. She also expressed her intention of entertaining those who had joined to lunch prior to their departure on Saturday, and further mentioned that the Rev H C Young and herself would provide motors for the journey to Taunton.

All seven of these recruits seem to have survived the First World War. None of them is named on the War Memorial adjacent to Church House.

Given the confusion during the early 1900s about who actually owned Church House, it is not surprising that the trustees wanted to avoid any future ambiguities. In 1935, the Lord's rent was extinguished with an agreement between the parties for a payment of the equivalent of 20 years of Lord's rent: 2 manors x 4d x 20 years = 13/4, or about £35 in today's money.

The War Office requisitioned Church House between 1940 and 1944 as a canteen for the soldiers billeted at Crowcombe Court who operated the searchlight batteries on the Quantock Hills. For a small village the soldiers' presence would have increased Crowcombe's population significantly, and no doubt contributed to the village's social life.

Above: During WWII, soldiers operated searchlight batteries in the Quantock Hills, above Crowcombe. (Mike Charles/Shutterstock)

Bottom: The Army's requisition of Church House in October 1940. (Somerset Heritage Centre)

After the war, Church House became a village hall again, where for some years, midday meals were served to Crowcombe village schoolchildren.

From 1959, Church House saw another series of physical changes to the building itself, including being connected to the mains sewerage that year. This adaption was no doubt a welcome relief for all who used Church House. A letter to the Somerset Education Office identifies the problems that Church House had traditionally suffered from:

> The present toilet facilities in this building consisting of only an Elsan closet, which is impermanent, unscreened and as the premises are regularly used by adults and children for school meals, the Trustees consider this facility to be most unsatisfactory ... The Local Authority are disturbed by the continuing nuisance of foul water from the existing sink (used for school meals), the drainage from which has no satisfactory disposal and is unhygienic.

Church House management committee must have been pleased when they succeeded with their application for funding for:

> a proper screened WC and Toilet, with a drainage connection to the new main sewer ... and a proper form of drainage also from the existing sink.

In 1963 the local authority built the car park at the rear and demolished the adjacent cottages (the poor houses) as part of a road widening scheme. Exposing the south-east elevation of Church House led to extensive structural work at that end of the building to make it sound once more.

Right: Letter to the builder requesting a late payment. (Somerset Heritage Centre)

These structural works included:

- parapet cappings and kneelers
- flashings for the parapet
- removal of plaster
- new stone work
- restoration of stairs landing to its original size, with new foundations
- balustrading and
- re-pointing. (29)

As ever, funds were short and the local authority decided initially that it could not afford a balustrade. However, six months later the architect was required to write to the builder:

> The Rector and General Slater on behalf of the Trustees have been observing the stone steps over the past weeks and have decided after all for safety reasons a balustrading should be provided before the visitors get busy...

Pressure was applied on the builder to accept late payment 'for the sake of the children who use the Hall for School Meals'. (Mar 1964)

The removal of the adjacent 'poor house' cottages exposed the end of Church House. This led to extensive structural work of the end wall and stairs. (Unknown source)

In 1981, a new roof was installed to the south east side. Then in 1995, a much-improved internal staircase was built. The original staircase installed as part of the 1908 restoration had posed great practical problems for hirers. It was very steep and covered by a hatch that required lifting to gain access to the gallery. There must have been many a dropped tray as people climbing the stairs tried to juggle what they were carrying while simultaneously opening the hatch above their heads.

Its 1995 successor was much less steep, requiring 15 steps to reach the top, as opposed to the original 12. It arrives at the gallery straightforwardly (without a hatch cover) at the north-west corner of the building, as shown in the floor plan. It was built by Jeff Duddridge's joinery firm, that for at least four generations has served the village's carpentry needs.

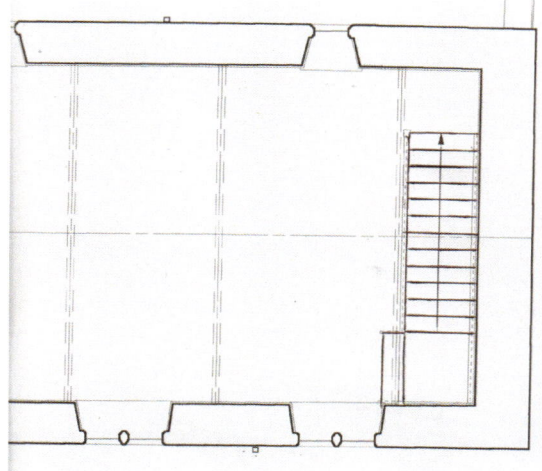

Above: *The 1995 replacement staircase improved internal access to the first floor. (Louise Crossman drawing, 2007)*

The aim had been to fit it with a stairlift. However, this enhancement that would have given access to the gallery for people with a disability, was vetoed by the Fire Service for safety reasons. The staircase would have become too narrow to be a satisfactory fire exit for the gallery.

In 1999, ownership of the pound, which is located at the western end of the building, was transferred from the Trollope-Bellew Estate to Church House. In its day, the pound was used to secure stray animals, animals grazing illegally on common lands or even to hold animals overnight for market. To this day, it remains a reminder of the village's market days.

In 2007, almost exactly 100 years after the Rev. Young's 1908 restoration, Church House underwent a major £80,000 internal refurbishment. With the aim of modernising the interior of what has become a local amenity available for hire, the work included:

- two WCs, including one for people with a disability
- oil central heating
- new lighting throughout, including specialist lighting for art exhibitions and
- comprehensive furnishing and equipping of the servery.

This substantial funding was achieved through a combination of local fundraising and grants from the Foyle Foundation, the Garfield Weston Foundation, Viridor Credits, Quantock Hills AONB, Somerset County Council, Somerset Rural Renaissance Partnership and West Somerset Council.

Having refurbished the interior of Church House, it was obvious that equivalent conservation was required to the exterior:

- 100 per cent lime mortar re-pointing of the entire building
- structural work to the parapet flashings and stone coping
- stitching of the masonry on the north-west corner of the building and
- new cast iron guttering and downpipes.

This proposal required raising another £64,000, achieved in 2016 through generous support from the village again and grants from the Heritage Lottery Fund, The Pilgrims Trust, Garfield Weston Foundation, Foyle Foundation and the Prince of Wales's Charitable Foundation.

Above: Church House covered in scaffolding mid project. (Paul Savage)

Top right: Stitching the masonry on the north-west corner of the building. (Paul Savage)

Bottom right: Preparing stone for the roof repairs. (Paul Savage)

The Heritage Lottery Fund also contributed to a range of activities for the community—including a panel on the history of Church House sited in the car park, a new leaflet, a new Church House website, this book and a lecture programme for groups of visitors from local history societies and schools.

Church House, sensitively restored to a high standard inside and out, is a magnificent venue available for hire — both to Crowcombe villagers and the wider West Somerset community.

Top left: The fully restored Church House, 2016. (Paul Savage)

Top right: Village party to celebrate the completion of the 2016 conservation project. (Catherine Brew)

Left: The gallery upstairs set out for a village event. (Catherine Brew)

Above: Annual flower show. (Paul Savage)

Opposite page:
Top: The downstairs meeting room. (Valori Menneer)

Middle and Bottom: The annual flower show. (Paul Savage)

REFERENCES

Substantial use was made of documents held by the Somerset Archives and Local Studies at Somerset Heritage Centre (SHC), whose staff we thank for their unstinting help and patience in assisting us with our enquiries. Reference numbers to their documents are shown at the end of each reference.

1. Clare Gathercole, *A brief history of Crowcombe*, Somerset Heritage Centre (SHC).

2. R.W. Dunning (ed.), *The Victoria history of the county of Somerset*, 1985.

3. John Collinson, *The history and antiquities of the county of Somerset,* 1791.

4. James Weir, *Church House Crowcombe,* 4 May 2016.

5. John Aubrey, *Miscellanies upon various subjects, London,* 1714.

6. E.H.D. Williams, 'Church Houses in Somerset', *Vernacular Architecture*, vol. 23, 1992.

7. William Greswell, 'The Crowcombe Church House, Somerset', *The Reliquary & Illustrated Archaeologist*, vol. 14, 1908.

8. M.L.R. Isaac, *Crowcombe Church House,* 1995.

9. Churchwardens' and overseers' accounts [D/P/Crow/13/2/6 & D/P/Crow 9/1/2, SHC].

10. Deeds and Rules of Crowcombe School 1733 [pDD/C132b, SHC].

11. Mary Rhodes, *Watchet*, Somerset, 2016.

12. Hilary Binding, *Exmoor by the way*, Halsgrove, 2009.

13. *West Somerset Gazette*, 26 September 1908.

14. Mrs Ethel Trollope to Charity Commissioners, 27 December 1905 [D/P/Crow/8/3/1, SHC].

15. The Rev. H. Christian Young to Charity Commissioners, 17 April 1906 [D/P/Crow/8/3/1, SHC].

16. The Rev. William Greswell to the Rev H. Christian Young, 31 March 1906 [D/P/Crow/8/3/1, SHC].

17. Charity Commission, 8 February 1907.

18. Charity Commissioners to Mrs Ethel Trollope, 10 May 1907 [D/P/Crow/8/3/1, SHC].

19. A. Basil Cottam to the Rev. H. Christian Young, 23 April 1907 [D/P/Crow/8/3/1, SHC].

20. A. Basil Cottam to the Rev. H. Christian Young, 4 May 1907 [D/P/Crow/8/3/2, SHC].

21. A. Basil Cottam to the Rev. H. Christian Young, 8 June 1907 [D/P/Crow/8/3/2, SHC].

22. A. Basil Cottam to Charity Commissioners, 23 July 1907 [D/P/Crow/8/3/1, SHC].

23. A. Basil Cottam to the Rev. H. Christian Young, 13 November 1907 [D/P/Crow/8/3/2, SHC].

24. A. Basil Cottam to the Rev. H. Christian Young, 28 January 1908 [D/P/Crow/8/3/2, SHC].

25. A. Basil Cottam to the Rev. H. Christian Young, 4 February 1908 [D/P/Crow/8/3/2, SHC].

26. Final accounts from A. Basil Cottam to Church House Trustees [1908, D/P/Crow/8/3/2, SHC].

27. *Somerset Free Press*, 26 September 1908.

28. *West Somerset Gazette*, 26 September 1908.

29. William Marsden, (Architect involved with removal of stairs) Specification, November 1963 [A/EAT/1/2/3, SHC].

30. Michael Richardson-Bunbury, *A Short History of Crowcombe*, Crowcombe Parish Council, 1999.

Acknowledgements

In putting this story together we have had all kinds of assistance, advice and contributions from many local and professional people, to all of whom we are indebted and are most grateful.

James Weir, the Historic Buildings Consultant, had visited Crowcombe's Church House some years earlier. As our fabric conservation project was getting underway, he generously returned to give a talk to the village on the historical significance of the building. He drew our attention to a number of key architectural features of the building that were totally unknown to us when we began on our journey.

Bob Croft, Head of Historic Environment at the Somerset Heritage Centre Trust, has been a regular source of encouragement and advice to us. We are equally most grateful to the staff of the Somerset Archives and Local Studies at the Somerset Heritage Centre for their unstinting help and patience with us - the lay archive sextet from our village: Sue Billinge, Geoffrey Billson, Priscilla Boddington, Gill Brown, Carolynne Fawcett and Peter Menneer.

We thank Valori Menneer, for her photography together with Paul Savage of R & P Photographic Service in the village and James Weir for his photos.

We are also most grateful for the professionalism and commitment of our architects, Claire Fear and Jenny Matravers of Architectural Thread Ltd of Wellington, Somerset. They drew our attention to all kinds of features in Church House's historical architecture that otherwise we would have missed.

The village of Crowcombe over the past two years have been amazingly supportive and generous to this conservation project. Through a combination of family donations, community organisation contributions and two fund raising events, this small village (comprising just over 200 families) raised very nearly £25,000. This considerable sum enabled the team, with heads held high, to apply confidently for grants from outside trusts and foundations.

Thank you too to our editors, Nigel Brew, Accredited Editor (IPEd), and Priscilla Boddington. Their expertise was most useful and helped us strive to make our text accurate, consistent and readable.

We have also had much appreciated help and involvement from a wide range of other villagers and local people - in many personal capacities: Alan Austin, Dan Davis, Jeff Duddridge, David Goodland, David Kenyon, Mairi MacKinnon, Mary Rhodes, Anthony Trollope-Bellew, Mark Wilson and:

- Naomi Cudmore of Exmoor Magazine
- Mark Lidster and Lydia Breeze-Chilcott from Corbel Conservation Ltd
- Liz Peeks of West Somerset Council,
- Steve McAuliffe of Webglu Ltd
- Josh Weddercopp from Crowcombe & Stogumber School

Above all we are most grateful to the Trusts and Foundations that have made financial contributions to the funding of the fabric conservation of this grade II* listed and much loved charity building:

- The Pilgrim Trust
- Garfield Weston Foundation
- Foyle Foundation
- The Prince of Wales's Charitable Foundation
- Heritage Lottery Fund - who in addition have funded a wide range of activities to achieve outcomes for people and the community, of which this book is one.

Finally we thank Peter Menneer's volunteer colleagues who have served on the management committee over this strenuous period: Chris Birch, Jonathan Denton, Maggie Inglis, Trish Kennedy, Jacqui MacQueen, Sheila Martin, Louise Normandale, Lynn Norris and our accountant Martin Smith - together with our Trustees:

- The Ven John Reed, Archdeacon of Taunton - recently retired and succeeded by the Ven Simon Hill
- The Rev Jon Rose, Rural Dean
- The Rev Angela Berners-Wilson, Rector
- Geoffrey Billson, Churchwarden
- Melville Trimble, Churchwarden

We have undertaken extensive research and fact-checking, making every effort to ensure the information held within is correct. We apologise for any errors that you may find.

Peter Menneer
Chairman of Church House Management Committee

Catherine Brew
Red Plait Interpretation LLP, Portland, Dorset